Ben Stacy Jerrik (Ed.)

Alabama State Route 123

Ben Stacy Jerrik (Ed.)

Alabama State Route 123

Alabama State Route 52, Alabama State Route 103, Fort Rucker, Alabama State Route 51, U.S. Route 84

Part Press

Contents

Articles

Alabama State Route 123	1
Dale County, Alabama	3
Houston County, Alabama	8
Geneva County, Alabama	12
Alabama State Route 52	16
Alabama State Route 103	18
Little Choctawhatchee River	20
Blackwell Field	21
Alabama State Route 51	23
U.S. Route 84	25
Alabama State Route 27	30
Newton, Alabama	32
Ozark, Alabama	35

References

Article Sources and Contributors	39
Image Sources, Licenses and Contributors	40

Alabama State Route 123

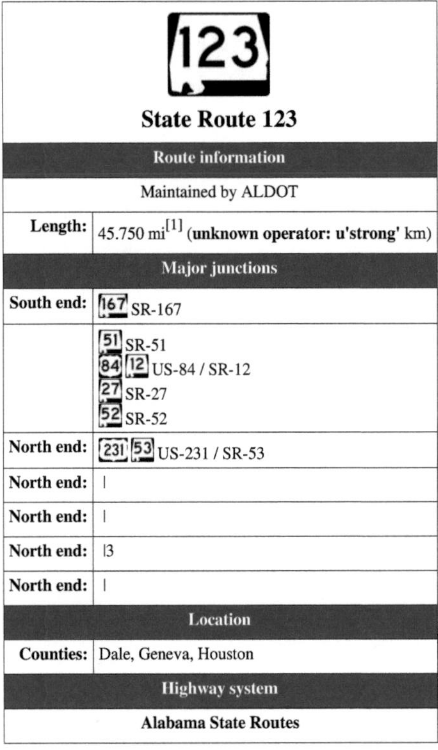

Length:	45.750 mi[1] (**unknown operator: u'strong'** km)
Major junctions	
South end:	SR-167
	SR-51 US-84 / SR-12 SR-27 SR-52
North end:	US-231 / SR-53
North end:	I
North end:	I
North end:	I3
North end:	I
Location	
Counties:	Dale, Geneva, Houston
Highway system	
Alabama State Routes	

State Route 123 or **SR-123** is a 45.750-mile (**unknown operator: u'strong'** km) state highway that serves Dale, Houston and Geneva Counties as a connection between Ariton, Ozark and Hartford. SR-123 intersects SR-167 at its southern terminus and US-231 at its northern terminus.

Route description

State Route 123 begins at its intersection with SR-167 just south of Hartford.[2] From this point, the route travels in northerly direction where it intersects SR-52 in the central business district of Hartford.[3] From Hartford, SR-123 travels in a northeasterly direction before turning to the north upon entering Houston County and intersecting SR-103.[4] The route continues through its intersection with US-84 and continues in its northerly course into Dale County after crossing the Little Choctawhatchee River.[5] SR-123 continues along its northerly course in Dale County where it intersects SR-134 and runs concurrent with it for approximately 2.5 miles (**unknown operator: u'strong'** km) through the town of Newton.[1] From the end of this concurrency through Ozark, SR-123 continues generally in this northern course in passing both Fort Rucker and Blackwell Field.[6] Upon leaving Ozark, the route continues in a northwesterly course in route to Ariton where it then turns in a westerly direction where it meets its northern terminus with US-231.[7]

Major intersections

County	Location	Mile[1]	Destinations	Notes
Geneva		0.0	SR-167	Southern terminus
	Hartford	1.424	SR-52	
Houston	Wicksburg	10.852	SR-103	
		12.854	US-84 / SR-12	
Dale	Newton	18.917	SR-134	Start of concurrency with SR-134
		21.345	SR-134	End of concurrency with SR-134
	Ozark	27.325	US-231 / SR-53	
		29.946	SR-27	
	Ariton	41.654	SR-51	Start of concurrency with SR-51
		42.095	SR-51	End of concurrency with SR-51
		45.750	US-231 / SR-53	Northern terminus

References

[1]　Alabama Department of Transportation (1999). *Milepost Map of Dale & Geneva Counties* (http://www.dot.state.al.us/NR/rdonlyres/ 7B15B300-359D-413A-A7FA-ED33C8C261CD/0/CountyMilePostMaps.pdf) (Map). Cartography by ALDOT Bureau of Transportation Planning, Survey & Mapping Division. .

[2]　Google, Inc. *Google Maps − SR-123 southern terminus* (http://maps.google.com/?ie=UTF8&ll=31.082948,-85.698327&spn=0,0. 009956&z=17&layer=c&cbll=31.082843,-85.698364&panoid=urCYaYGZO6XqOm08jIIc7A&cbp=12,262.04,,0,9.49) (Map). Cartography by Google, Inc. . Retrieved April 29, 2010.

[3]　Google, Inc. *Google Maps − SR-123/ SR-52 junction* (http://maps.google.com/?ie=UTF8&ll=31.10237,-85.695462&spn=0.012016,0. 019913&z=16) (Map). Cartography by Google, Inc. . Retrieved April 29, 2010.

[4]　Google, Inc. *Google Maps − SR-123/ SR-103 junction* (http://maps.google.com/?ie=UTF8&ll=31.213517,-85.627946&spn=0,0. 009956&z=17&layer=c&cbll=31.213591,-85.627975&panoid=XjGDPpsZdBFkWhFCKaTGkg&cbp=12,26.17,,0,2.19) (Map). Cartography by Google, Inc. . Retrieved April 29, 2010.

[5]　Google, Inc. *Google Maps − SR-123/ US-84 junction* (http://maps.google.com/?ie=UTF8&ll=31.259697,-85.625639&spn=0.047984,0. 079651&z=14) (Map). Cartography by Google, Inc. . Retrieved April 29, 2010.

[6]　Google, Inc. *Google Maps − SR-123 between SR-134 and Ozark* (http://maps.google.com/?ie=UTF8&ll=31.409619,-85.608902& spn=0.190459,0.323753&z=12) (Map). Cartography by Google, Inc. . Retrieved April 29, 2010.

[7]　Google, Inc. *Google Maps − SR-123 between Ozark & US-231* (http://maps.google.com/?ie=UTF8&ll=31.536993,-85.687523&spn=0. 1902,0.323753&z=12) (Map). Cartography by Google, Inc.. Retrieved April 29, 2010.

Dale County, Alabama

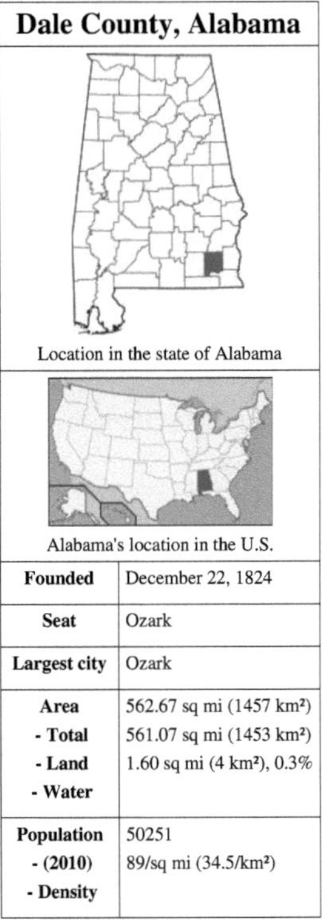

Dale County, Alabama

Location in the state of Alabama

Alabama's location in the U.S.

Founded	December 22, 1824
Seat	Ozark
Largest city	Ozark
Area - Total - Land - Water	562.67 sq mi (1457 km²) 561.07 sq mi (1453 km²) 1.60 sq mi (4 km²), 0.3%
Population - (2010) - Density	50251 89/sq mi (34.5/km²)

Dale County is a county of the U.S. state of Alabama. The vast majority of Fort Rucker U.S. Army Aviation Center for Excellence is located in Dale County. Its name is in honor of General Samuel Dale. As of the 2010 census the population was 50,251. Its county seat and largest city is Ozark.

Dale County is part of the Enterprise–Ozark Micropolitan Statistical Area.

History

The area now known as Dale County was originally inhabited by members of the Creek Indian nation, who occupied all of southeastern Alabama during this period. The county, together with the surrounding area, was ceded to the United States in the 1814 Treaty of Fort Jackson, ending the Creek Indian Wars. A blockhouse had been constructed during the conflict on the northwestern side of the Choctawhatchee River, and the first non-Indian residents of Dale County would be veterans who began to settle in the area around 1820.[1]

Dale County was established on December 22, 1824. It originally included the whole of what is now Coffee County and Geneva County, together with the "panhandle" portion of Houston County. The original county seat was located at Dale's Court House (now the town of Daleville), but when Coffee County split from Dale in 1841, the seat was moved to Newton. Here it remained until 1870 when, following a courthouse fire in 1869 and the formation of Geneva County (which took the southern third of Dale County), the county seat was moved to the town of Ozark, where it remains. In 1903 a small portion of the southeast part of Dale county was joined to the newly-formed

Houston County.

Portions of the 15th Regiment of Alabama Infantry, which served with great distinction throughout the U.S. Civil War, were recruited in Dale County, with all of Co. "E" and part of Co. "H" being composed of Dale County residents. This unit is most famous for being the regiment that confronted the 20th Maine on the Little Round Top during the Battle of Gettysburg on July 2, 1863. Despite several ferocious assaults, the 15th was ultimately unable to dislodge the Union troops, and was ultimately forced to retreat after a desperate bayonet charge led by the 20th Maine's commander, Col. Joshua L. Chamberlain.[2] This assault was vividly recreated in Ronald F. Maxwell's 1993 film *Gettysburg*.

Geography

Dale County is located in the southeastern corner of the State of Alabama. According to the 2000 census, the county has a total area of 562.67 square miles (**unknown operator: u'strong'** km^2), of which 561.07 square miles (**unknown operator: u'strong'** km^2) (or 99.72%) is land and 1.60 square miles (**unknown operator: u'strong'** km^2) (or 0.28%) is water.[3]

Major highways

- 84 U.S. Highway 84
- 231 U.S. Highway 231
- 27 State Route 27
- 51 State Route 51
- 85 State Route 85
- 92 State Route 92
- 123 State Route 123
- 134 State Route 134

Adjacent counties

- Barbour County (north)
- Henry County (east)
- Houston County (southeast)
- Geneva County (southwest)
- Coffee County (west)
- Pike County (northwest)

Demographics

Dale County, Alabama		
Year	Pop.	±%
1830	2031	—
1840	7397	+264.2%
1850	6382	–13.7%
1860	12197	+91.1%
1870	11325	–7.1%
1880	12677	+11.9%
1890	17225	+35.9%
1900	21189	+23.0%
1910	21608	+2.0%
1920	22711	+5.1%
1930	23175	+2.0%
1940	22685	–2.1%
1950	20828	–8.2%
1960	31066	+49.2%
1970	52995	+70.6%
1980	47821	–9.8%
1990	49633	+3.8%
2000	49129	–1.0%
2010	50251	+2.3%

Sources: "American FactFinder" [4]. United States Census Bureau. through 1960 [5]

As of the census[6] of 2000, there were 49,129 people, 18,878 households, and 13,629 families residing in the county. The population density was 88 people per square mile (34/km^2). There were 21,779 housing units at an average density of 39 per square mile (15/km^2). The racial makeup of the county was 74.4% White, 20.4% Black or African American, 0.60% Native American, 1.1% Asian, 0.15% Pacific Islander, 1.3% from other races, and 2.2% from two or more races. 3.4% of the population were Hispanic or Latino of any race. 2.85% of the population reported speaking Spanish at home, while 1.51% speak German.[7]

There were 18,878 households out of which 36% had children under the age of 18 living with them, 55% were married couples living together, 13.6% had a female householder with no husband present, and 27.8% were non-families. 24.3% of all households were made up of individuals and 8.8% had someone living alone who was 65 years of age or older. The average household size was 2.5 and the average family size was 3.0.

In the county the population was spread out with 26.6% under the age of 18, 9.6% from 18 to 24, 30.3% from 25 to 44, 21.8% from 45 to 64, and 11.8% who were 65 years of age or older. The median age was 34 years. For every 100 females there were 98.3 males. For every 100 females age 18 and over, there were 95 males.

The median income for a household in the county was $31,998, and the median income for a family was $37,806. Males had a median income of $29,844 versus $19,988 for females. The per capita income for the county was $16,010. 15% of the population and 12.6% of families were below the poverty line. 19.4% of those under the age of 18 and 16.5% of those 65 and older were living below the poverty line.

Municipalities and census-designated places

- Ariton
- Asbury (northeast of Ozark)
- Clayhatchee
- Daleville
- Dothan (part - most of Dothan is in Houston County, with additional portions in Henry County)
- Enterprise (part - most of Enterprise is in Coffee County)
- Fort Rucker (U.S. Army base, treated as a census-designated place)
- Grimes
- Level Plains
- Midland City
- Napier Field
- Newton
- Ozark
- Pinckard

Unincorporated communities

- Arguta (southwest of Blue Springs)
- Barefield Crossroads (west of Abbeville)
- Barnes (southeast of Abbeville)
- Beamon (northeast of Ozark)
- Bells Crossroads (northwest of Clayton)
- Bertha (southeast of Blue Springs)
- Bethel (southwest of Clio)
- Browns Crossroad (northeast of Newton)
- Clopton (southeast of Blue Springs)
- Dill (north of Ozark)
- Dillard (northwest of Ozark)
- Dykes Crossroad (southwest of Blue Springs)
- Echo (east of Ozark)
- Ewell (southeast of Ozark)
- Gerald (northwest of Daleville)
- Kelly (southwest of Newton)
- Mabson (east of Ozark)
- Marley Mill (northwest of Ozark)
- Plainview (north of Newton)
- Roberts Crossroads (south of Blue Springs)
- Rocky Head (southwest of Clio)
- Skipperville (northeast of Ozark)
- Sylvan Grove (northeast of Midland City)
- Waterford (north of Newton)

See also

- National Register of Historic Places listings in Dale County, Alabama
- Properties on the Alabama Register of Landmarks and Heritage in Dale County, Alabama

References

[1] http://web.archive.org/web/20080528150204/http://www.geocities.com/Heartland/Prairie/1767/.Retrieved on 21 July 2008.

[2] Desjardin, pp. 69-71, Pfanz, p. 232.

[3] "Census 2000 U.S. Gazetteer Files: Counties" (http://www.census.gov/tiger/tms/gazetteer/county2k.txt). United States Census. . Retrieved 2011-02-13.

[4] http://factfinder2.census.gov/faces/nav/jsf/pages/index.xhtml

[5] http://fisher.lib.virginia.edu/collections/stats/histcensus/php/newlong.php?subject=1

[6] "American FactFinder" (http://factfinder.census.gov). United States Census Bureau. . Retrieved 2008-01-31.

[7] http://www.mla.org/map_data_results&state_id=1&county_id=45&mode=geographic&zip=&place_id=&cty_id=&ll=&a=&ea=& order=r

Houston County, Alabama

<table>
<tr><td colspan="2" align="center">Houston County, Alabama</td></tr>
<tr><td colspan="2" align="center">
Houston County courthouse in Dothan, Alabama</td></tr>
<tr><td colspan="2" align="center">
Location in the state of Alabama</td></tr>
<tr><td colspan="2" align="center">
Alabama's location in the U.S.</td></tr>
<tr><td>Founded</td><td>February 9, 1903</td></tr>
<tr><td>Seat</td><td>Dothan</td></tr>
<tr><td>Largest city</td><td>Dothan</td></tr>
<tr><td>Area
- Total
- Land
- Water</td><td>581.65 sq mi (1506 km²)
580.36 sq mi (1503 km²)
1.29 sq mi (3 km²), 0.22%</td></tr>
<tr><td>Population
- (2010)
- Density</td><td>101547
175/sq mi (67.5/km²)</td></tr>
<tr><td>Time zone</td><td>Central: UTC-6/-5</td></tr>
</table>

Houston County is a county of the U.S. state of Alabama. As of 2010 the population was 101,547. Its county seat is Dothan.

Houston County is part of the Dothan Metropolitan Statistical Area.

History

Houston County was established on February 9, 1903, from parts of Dale County, Geneva County and Henry County. It was named after George Smith Houston, a Governor of Alabama.

Government

As of 2011, the County Chairman is held by Mark Culver, while William Dempsey is the Chief Administrative Officer. The County Commission is District 1, Curtis Harvey; District 2, Doug Sinquefield; District 3, Jackie Battles; and District 4, Phil Forrester. Andy Hughes serves as Sheriff, Probate Judge is Luke Cooley, Revenue Commissioner is Starla Moss Matthews. Houston County is located in Alabama's Second Congressional District; its current Representative (as of 2011) is Martha Roby (R).

Geography

According to the 2000 census, the county has a total area of 581.65 square miles (**unknown operator: u'strong'** km^2), of which 580.36 square miles (**unknown operator: u'strong'** km^2) (or 99.78%) is land and 1.29 square miles (**unknown operator: u'strong'** km^2) (or 0.22%) is water.[1]

Major highways

- 84 U.S. Highway 84
- 231 U.S. Highway 231
- 431 U.S. Highway 431
- 52 State Route 52
- 53 State Route 53
- 95 State Route 95

Adjacent counties

- Henry County (north)
- Early County, Georgia (east)
- Seminole County, Georgia (southeast)
- Jackson County, Florida (south)
- Geneva County (west)
- Dale County (northwest)

Demographics

Houston County, Alabama		
Year	Pop.	±%
1910	32414	—
1920	37334	+15.2%
1930	45935	+23.0%
1940	45665	−0.6%
1950	46522	+1.9%
1960	50718	+9.0%
1970	56574	+11.5%
1980	74632	+31.9%
1990	81331	+9.0%
2000	88787	+9.2%
2010	101547	+14.4%
Sources: "American FactFinder" [4]. United States Census Bureau. through 1960 [5]		

As of the census[2] of 2000, there were 88,787 people, 35,834 households, and 25,119 families residing in the county. The population density was 153 people per square mile (59/km^2). There were 39,571 housing units at an average density of 68 per square mile (26/km^2). The racial makeup of the county was 73.08% White, 24.60% Black or African American, 0.37% Native American, 0.62% Asian, 0.02% Pacific Islander, 0.39% from other races, and 0.92% from two or more races. 1.26% of the population were Hispanic or Latino of any race.

There were 35,834 households out of which 33.00% had children under the age of 18 living with them, 52.50% were married couples living together, 14.10% had a female householder with no husband present, and 29.90% were non-families. 26.40% of all households were made up of individuals and 10.10% had someone living alone who was 65 years of age or older. The average household size was 2.45 and the average family size was 2.95.

In the county the population was spread out with 25.90% under the age of 18, 8.20% from 18 to 24, 28.70% from 25 to 44, 23.50% from 45 to 64, and 13.70% who were 65 years of age or older. The median age was 37 years. For every 100 females there were 95.0 males. For every 100 females age 18 and over, there were 86.20 males.

The median income for a household in the county was $34,431, and the median income for a family was $42,437. Males had a median income of $32,092 versus $21,409 for females. The per capita income for the county was $18,759. About 11.80% of families and 15.00% of the population were below the poverty line, including 21.10% of those under age 18 and 16.30% of those age 65 or over.

Cities and towns

- Ashford
- Avon
- Columbia
- Cottonwood
- Cowarts
- Dothan (parts of Dothan are in Dale County and in Henry County)
- Gordon
- Kinsey
- Madrid

- Rehobeth
- Taylor
- Webb

See also

- National Register of Historic Places listings in Houston County, Alabama
- Properties on the Alabama Register of Landmarks and Heritage in Houston County, Alabama

References

[1] "Census 2000 U.S. Gazetteer Files: Counties" (http://www.census.gov/tiger/tms/gazetteer/county2k.txt). United States Census. . Retrieved 2011-02-13.
[2] "American FactFinder" (http://factfinder.census.gov). United States Census Bureau. . Retrieved 2008-01-31.

External links

- Houston County web site (http://www.houstoncounty.org/)

Geneva County, Alabama

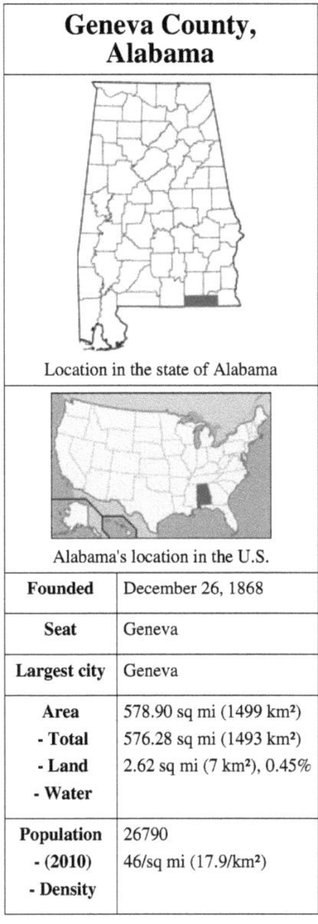

Geneva County, Alabama	
Location in the state of Alabama	
Alabama's location in the U.S.	
Founded	December 26, 1868
Seat	Geneva
Largest city	Geneva
Area	578.90 sq mi (1499 km²)
- Total	576.28 sq mi (1493 km²)
- Land	2.62 sq mi (7 km²), 0.45%
- Water	
Population	26790
- (2010)	46/sq mi (17.9/km²)
- Density	

Geneva County is a county of the U.S. state of Alabama. As of 2010 the population was 26,790. Its county seat is Geneva. Geneva County is a prohibition or dry county.

The county was named after its county seat, which in turn was named after Geneva, New York which was named after Geneva, Switzerland, by Walter H. Yonge, an early town resident and Swiss native.[1]

Geneva County is part of the Dothan, Alabama, Metropolitan Statistical Area.

History

- Geneva County was established on December 26, 1868.
- The county was declared a disaster area in September 1979 due to damage from Hurricane Frederic.
- On March 10, 2009, a gunman named Michael McLendon went on a mass shooting spree at 9 locations in Geneva County from the town of Samson to the city of Geneva killing 11 people and wounding over 15. McLendon entered his former place of employment, Reliable Metal Products on the northeastern side of Geneva, where he took his own life.

Geography

According to the 2000 census, the county has a total area of 578.90 square miles (**unknown operator: u'strong'** km^2), of which 576.28 square miles (**unknown operator: u'strong'** km^2) (or 99.55%) is land and 2.62 square miles (**unknown operator: u'strong'** km^2) (or 0.45%) is water.[2]

Major highways

- 27 State Route 27
- 52 State Route 52
- 54 State Route 54
- 85 State Route 85
- 87 State Route 87

Adjacent counties

- Dale County (north-northeast)
- Houston County (east)
- Holmes County, Florida (south)
- Walton County, Florida (southwest)
- Covington County (west)
- Coffee County (north-northwest)

Demographics

Geneva County, Alabama		
Year	Pop.	±%
1870	2959	—
1880	4342	+46.7%
1890	10690	+146.2%
1900	19096	+78.6%
1910	26230	+37.4%
1920	29315	+11.8%
1930	30104	+2.7%
1940	29172	−3.1%
1950	25899	−11.2%
1960	22310	−13.9%
1970	21924	−1.7%
1980	24253	+10.6%
1990	23647	−2.5%
2000	25764	+9.0%
2010	26790	+4.0%

Sources: "American FactFinder" [4]. United States Census Bureau. through 1960 [5]

2010

Whereas according to the 2010 U.S. Census Bureau:

- 86.3% White
- 9.5% Black
- 0.8% Native American
- 0.3% Asian
- 0.0% Native Hawaiian or Pacific Islander
- 1.6% Two or more races
- 3.4% Hispanic or Latino (of any race)

2000

As of the census[3] of 2000, there were 25,764 people, 10,477 households, and 7,459 families residing in the county. The population density was 45 people per square mile ($17/km^2$). There were 12,115 housing units at an average density of 21 per square mile ($8/km^2$). The racial makeup of the county was 87.11% White, 10.65% Black or African American, 0.76% Native American, 0.12% Asian, 0.02% Pacific Islander, 0.62% from other races, and 0.72% from two or more races. 1.76% of the population were Hispanic or Latino of any race.

There were 10,477 households out of which 30.60% had children under the age of 18 living with them, 56.40% were married couples living together, 11.00% had a female householder with no husband present, and 28.80% were non-families. 26.30% of all households were made up of individuals and 12.30% had someone living alone who was 65 years of age or older. The average household size was 2.43 and the average family size was 2.92.

In the county the population was spread out with 24.00% under the age of 18, 7.50% from 18 to 24, 26.80% from 25 to 44, 25.30% from 45 to 64, and 16.30% who were 65 years of age or older. The median age was 39 years. For every 100 females there were 94.70 males. For every 100 females age 18 and over, there were 90.00 males.

The median income for a household in the county was $26,448, and the median income for a family was $32,563. Males had a median income of $26,018 versus $19,341 for females. The per capita income for the county was $14,620. About 15.90% of families and 19.60% of the population were below the poverty line, including 27.20% of those under age 18 and 21.80% of those age 65 or over.

Cities and towns

- Black
- Coffee Springs
- Geneva
- Hartford
- Malvern
- Samson
- Slocomb
- Taylor

Unincorporated Communities

- Bailey Crossroads (north of Slocomb)
- Bald Hill
- Bellwood (east of Coffee Springs)
- Chancellor (northeast of Coffee Springs)
- Dundee (northeast of Hartford)
- Earlytown
- Eunola (northeast of Geneva)
- Fadette
- Hacoda (southwest of Earlytown)
- High Bluff (northwest of Hartford)
- Highfalls
- Highnote
- Kellys Crossroads

See also

- National Register of Historic Places listings in Geneva County, Alabama
- Properties on the Alabama Register of Landmarks and Heritage in Geneva County, Alabama

References

[1] "Geneva County" (http://www.archives.state.al.us/counties/geneva.html). Alabama Department of Archives and History. June 4, 2009. . Retrieved Aug 1, 2009.

[2] "Census 2000 U.S. Gazetteer Files: Counties" (http://www.census.gov/tiger/tms/gazetteer/county2k.txt). United States Census. . Retrieved 2011-02-13.

[3] "American FactFinder" (http://factfinder.census.gov). United States Census Bureau. . Retrieved 2008-01-31.

External links

- Official Site (http://www.genevacounty.net)

Alabama State Route 52

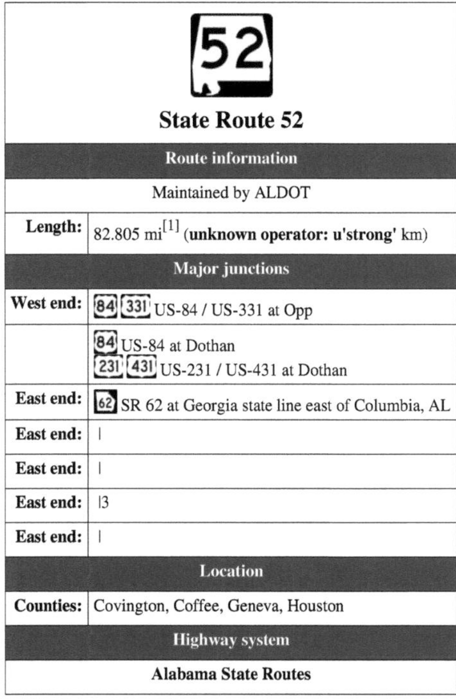

	State Route 52
	Route information
	Maintained by ALDOT
Length:	82.805 mi[1] (**unknown operator: u'strong'** km)
	Major junctions
West end:	84 331 US-84 / US-331 at Opp
	84 US-84 at Dothan 231 431 US-231 / US-431 at Dothan
East end:	62 SR 62 at Georgia state line east of Columbia, AL
East end:	I
East end:	I
East end:	I3
East end:	I
	Location
Counties:	Covington, Coffee, Geneva, Houston
	Highway system
	Alabama State Routes

State Route 52 is an 84-mile (**unknown operator: u'strong'** km) long route in the southeastern part of the state. The western terminus of the route is at a junction with U.S. Highway 331 at Opp. The eastern terminus of the route is at the Georgia state line east of Columbia. After State Route 52 crosses the Chattahoochee River and enters Georgia, the route continues as Georgia State Route 62.

Route description

State Route 52 serves as a parallel route of U.S. Highway 84. Beginning at Opp, State Route 52 assumes a southeastward trajectory as it leads towards Samson in western Geneva County. At Slocomb, the route briefly turns to the east before resuming its southeastward trajectory as it approaches the town of Geneva.

East of Geneva, State Route 52 assumes a slight northeastward trajectory. It passes through Hartford, the birthplace of Baseball Hall of Fame member Early Wynn, then continues through Slocomb as it leads towards Dothan, the largest city of Alabama's Wiregrass Region.

State Route 52 has junctions with three U.S. Highways at Dothan. On the city's west side, the route meets U.S. Highway 231. On the east side of Dothan, State Route 52 junctions U.S. Highways 84 and 431. All three U.S. Highways are aligned along Ross Clark Circle, a circumferential bypass of Dothan. As State Route 52 leads eastward out of Dothan, it passes through rural areas of Houston County as it leads towards Columbia and the Georgia state line.

Major intersections

County	Location	Mile[1]	Destinations	Notes
Covington	Opp	0.000	**84** **331** US-84 (SR-12) / US-331 (SR-9)	
Coffee		6.507	**189** SR-189 north	
Geneva		14.155	**54** SR-54 west	
		14.754	**153** SR-153 south	
	Samson	13.621	**87** SR-87	
	Geneva	28.742	**196** SR-196 east	
		30.564	**27** SR-27 south	West end of SR-27 overlap
		31.009	**27** SR-27 north	East end of SR-27 overlap
	Hartford	41.101	**167** SR-167	
		41.910	**123** SR-123	
	Slocomb	48.072	**103** SR-103 north	West end of SR-103 overlap
		48.149	**103** SR-103 south	East end of SR-103 overlap
Houston	Dothan	60.870	**210** SR-210	
		62.545	**84** US-84 west (SR-12)	West end of US-84 overlap
		63.711	**231** **431** US-231 (SR-53) / US-431 (SR-1)	
		65.136	**84** US-84 east (SR-12)	East end of US-84 overlap
		65.737	**210** SR-210	
	Columbia	81.806	**95** SR-95 south	West end of SR-95 overlap
		81.806	**95** SR-95 north	East end of SR-95 overlap
Chattahoochee River		82.805	Alabama–Georgia state line	
Early			**62** SR 62 east	Continuation into Georgia
1.000 mi = 1.609 km; 1.000 km = 0.621 mi				
Concurrency terminus • Closed/former • Incomplete access • Unopened				

References

[1] Alabama Department of Transportation. "Milepost Maps" (http://aldotgis.dot.state.al.us/milepostinternet/default.aspx). . Retrieved August 27, 2011.

Alabama State Route 103

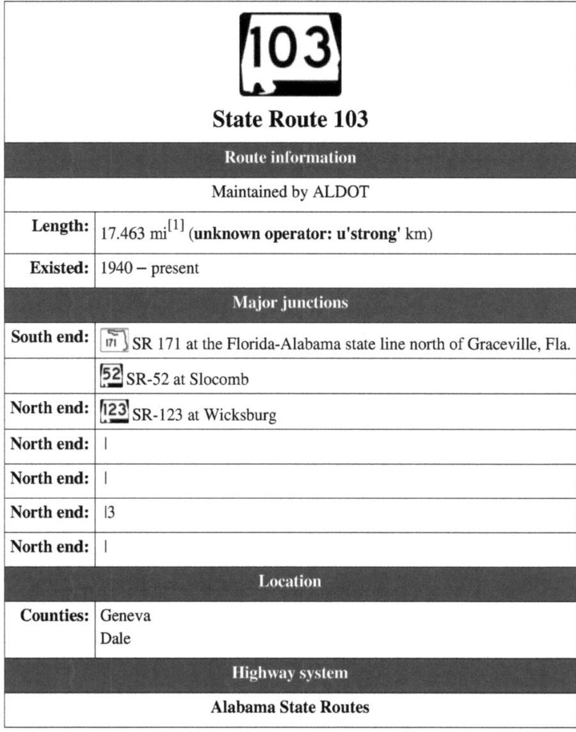

State Route 103

Route information	
Maintained by ALDOT	
Length:	17.463 mi[1] (**unknown operator: u'strong'** km)
Existed:	1940 – present
Major junctions	
South end:	SR 171 at the Florida-Alabama state line north of Graceville, Fla.
	SR-52 at Slocomb
North end:	SR-123 at Wicksburg
North end:	I
North end:	I
North end:	I3
North end:	I
Location	
Counties:	Geneva Dale
Highway system	
Alabama State Routes	

State Route 103 is a 17-mile (**unknown operator: u'strong'** km) long route in southeastern Alabama. The route begins at the Florida-Alabama state line and is a continuation of Florida State Road 171. The northern terminus of the route is at its junction with State Route 123 at Wicksburg, an unincorporated community north of the Geneva-Dale County line.

Route description

State Route 103 travels primarily through eastern Geneva County. It is one of numerous routes that enter Alabama from the Florida Panhandle. Upon entering Alabama, the route leads northward, passing primarily through rural areas of the county along a two-lane roadway. At the approximate halfway point of the route, it passes through Slocomb, the only incorporated community along its route.

Major intersections

County	Location	Mile[1]	Destinations	Notes
Geneva		0.000	SR 171	
	Slocomb	8.933	SR-52	
Houston	Wicksburg	17.463	SR-123	
1.000 mi = 1.609 km; 1.000 km = 0.621 mi				

References

[1] Alabama Department of Transportation. "Milepost Maps" (http://aldotgis.dot.state.al.us/milepostinternet/default.aspx). . Retrieved June 18, 2011.

Little Choctawhatchee River

Little Choctawhatchee River

Little Choctawhatchee River is a 24.0-mile-long (**unknown operator: u'strong'** km)[1] river in Alabama, United States. It drains an area of 261 square miles (**unknown operator: u'strong'** km^2) in Dale, Geneva, Henry and Houston counties. It empties into the Choctawhatchee River.

Surveys of the river show it to be poor in invertebrates and high in pollutants.[2]

References

[1] U.S. Geological Survey. National Hydrography Dataset high-resolution flowline data. The National Map (http://viewer.nationalmap.gov/ viewer/), accessed April 15, 2011

[2] "Nonpoint Source Screening Assessment of Southeast Alabama River Basins – 1999: CHOCTAWHATCHEE RIVER BASIN" (http:// www.adem.state.al.us/FieldOps/WQReports/NPSScAssesChoctawhatchee.pdf) (pdf). *Alabama Department of Environmental Management*. 2004-04-05. .

Blackwell Field

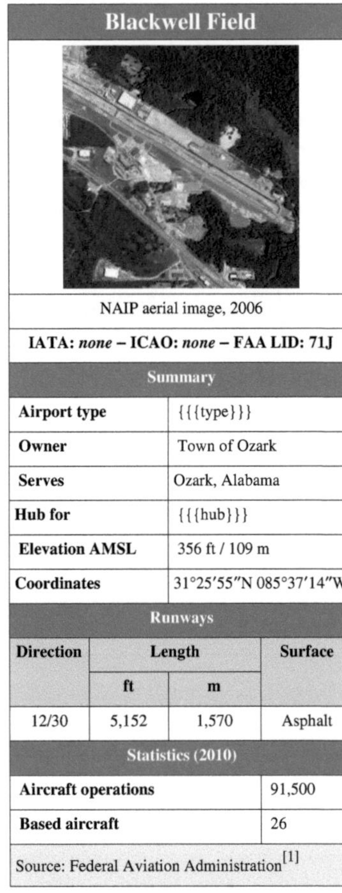

Blackwell Field

NAIP aerial image, 2006

IATA: *none* – **ICAO:** *none* – **FAA LID: 71J**

Summary	
Airport type	{{{type}}}
Owner	Town of Ozark
Serves	Ozark, Alabama
Hub for	{{{hub}}}
Elevation AMSL	356 ft / 109 m
Coordinates	31°25′55″N 085°37′14″W

Runways			
Direction	**Length**		**Surface**
	ft	**m**	
12/30	5,152	1,570	Asphalt

Statistics (2010)	
Aircraft operations	91,500
Based aircraft	26

Source: Federal Aviation Administration[1]

Blackwell Field (FAA LID: **71J**) is a public-use airport located two nautical miles (2.3 mi, 3.7 km) southeast of the central business district of Ozark, in Dale County, Alabama, United States. The airport is owned by the Town of Ozark.[1] It is included in the FAA's National Plan of Integrated Airport Systems for 2011–2015, which categorized it as a *general aviation* facility.[2]

Facilities and aircraft

Blackwell Field covers an area of 114 acres (**unknown operator: u'strong'** ha) at an elevation of 356 feet (109 m) above mean sea level. It has one runway designated 12/30 with an asphalt surface measuring 5,152 by 80 feet (1,570 x 24 m).[1]

For the 12-month period ending March 4, 2010, the airport had 91,500 general aviation aircraft operations, an average of 250 per day. At that time there were 26 aircraft based at this airport: 58% single-engine, 15% multi-engine, 4% jet and 23% helicopter.[1]

References

[1] FAA Airport Master Record for 71J (http://www.gcr1.com/5010web/airport.cfm?Site=71J) (Form 5010 (http://www.gcr1.com/ 5010web/Rpt_5010.asp?au=PU&o=PU&faasite=00480.1*A&fn=71J) PDF). Federal Aviation Administration. Effective 30 June 2011.

[2] National Plan of Integrated Airport Systems (http://www.faa.gov/airports/planning_capacity/npias/) for 2011–2015: Appendix A (PDF, 2.03 MB) (http://www.faa.gov/airports/planning_capacity/npias/reports/media/2011/npias_2011_appA.pdf). Federal Aviation Administration. Updated 4 October 2010.

External links

- FAA Terminal Procedures for 71J (http://aeronav.faa.gov/digital_tpp_search.asp?fldIdent=71J& fld_ident_type=FAA&ver=1203&eff=03-08-2012&end=04-05-2012&submit1=Search), effective 8 March 2012
- Resources for this airport:
    - AirNav airport information for 71J (http://www.airnav.com/airport/71J)
    - FlightAware airport information (http://flightaware.com/resources/airport/71J) and live flight tracker (http:/ /flightaware.com/live/airport/71J)
    - SkyVector aeronautical chart for 71J (http://skyvector.com/perl/code?id=71J&scale=2)

Alabama State Route 51

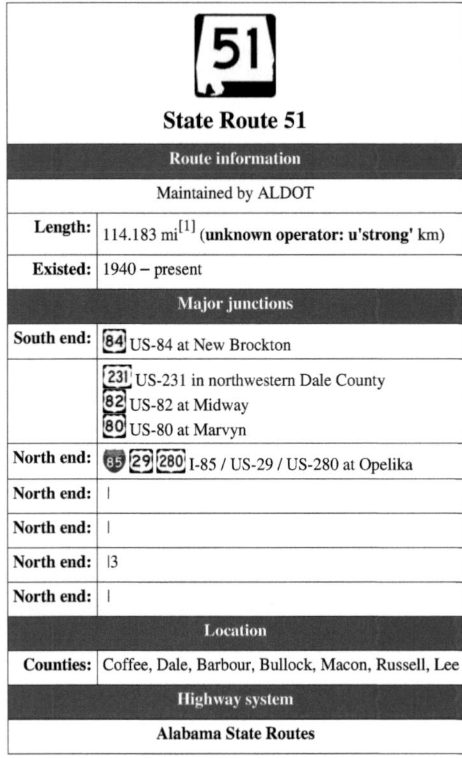

State Route 51

Route information	
Maintained by ALDOT	
Length:	114.183 mi[1] (**unknown operator: u'strong'** km)
Existed:	1940 – present
Major junctions	
South end:	[84] US-84 at New Brockton
	[231] US-231 in northwestern Dale County [82] US-82 at Midway [80] US-80 at Marvyn
North end:	[85] [29] [280] I-85 / US-29 / US-280 at Opelika
North end:	I
North end:	I
North end:	I3
North end:	I
Location	
Counties:	Coffee, Dale, Barbour, Bullock, Macon, Russell, Lee
Highway system	
Alabama State Routes	

State Route 51 is a 115-mile (**unknown operator: u'strong'** km) long route in the southeastern and east-central part of the state. The southern terminus of the route is at its junction with U.S. Highway 84 near New Brockton. The northern terminus of the route is at an interchange with Interstate 85 and U.S. Highways 29 and 280 at Opelika.

Route description

While it is signed as a north–south route, the orientation of State Route 51 is rather irregular. From its beginning near New Brockton, the route heads in a northeasterly direction, passing through rural areas and small towns in the southeastern part of the state.

State Route 51 passes through Clio, the birthplace of former Alabama governor George C. Wallace and Baseball Hall of Fame member Don Sutton. At Clio, the route turns briefly turns northward, then resumes its northward trajectory as it leads towards Clayton. At Clayton, the route then turns northwestward then northward as it leads towards Midway.

At Midway, State Route 51 joins U.S. Highway 82 and turns westward. The concurrency of the two routes ends east of Union Springs. State Route 51 then resumes a northeastward trajectory until it approaches Hurtsboro. The route then turns northward, continuing this orientation until it reaches its terminus at Opelika.

When State Route 51 was created in 1940, it covered only the route between Clayton and Midway. The route was extended to its current southern terminus in 1957, and was extended to its current northern terminus in 1986.

Major intersections

County	Location	Mile	Destinations	Notes
Coffee	Enterprise	0.000	US-84 (SR-12)	
		4.427	SR-122 south	
		5.779	SR-167 south	South end of SR-167 overlap
		5.888	SR-167 north	North end of SR-167 overlap
Dale		19.254	US-231 (SR-53)	
	Ariton	22.725	SR-123 west	South end of SR-123 overlap
		23.167	SR-123 east	North end of SR-123 overlap
Barbour	Clio	33.584	SR-10	
		39.738	SR-130 west	
	Clayton	48.695	SR-30 east	
		49.064	SR-239	
		49.933	SR-239	
Bullock	Midway	64.901	US-84 east (SR-6)	South end of US-84 overlap
		71.560	US-84 west (SR-6)	
Macon	No major junctions			
Russell	Hurtsboro	85.781	SR-26 east	
Lee	Marvyn	100.356	US-80 (SR-8)	
	Opelika	113.598	SR-169 south	
		114.183	I-85 / US-29 (SR-15) / US-280 (SR-38)	
1.000 mi = 1.609 km; 1.000 km = 0.621 mi				
Concurrency terminus • Closed/former • Incomplete access • Unopened				

References

[1] Alabama Department of Transportation. "Milepost Maps" (http://aldotgis.dot.state.al.us/milepostinternet/default.aspx). . Retrieved June 29, 2011.

U.S. Route 84

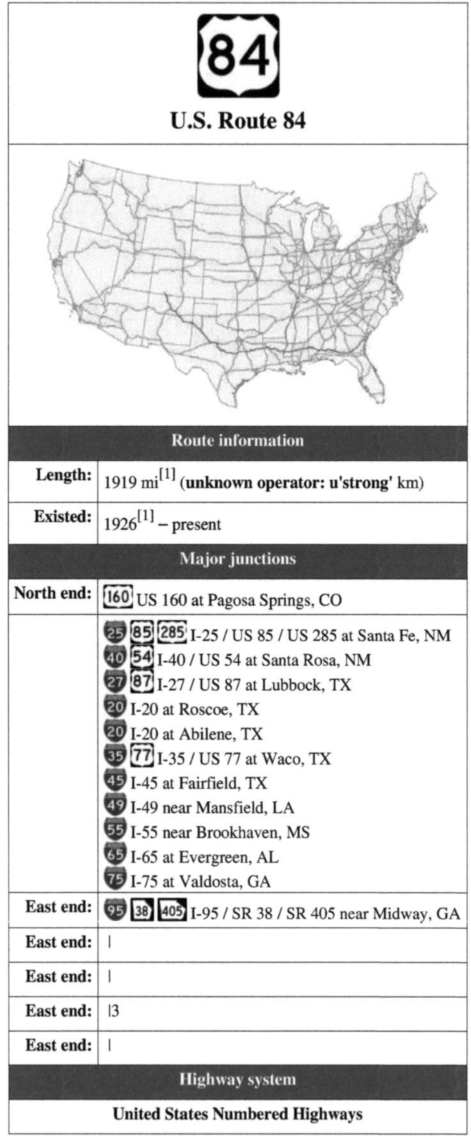

Route information	
Length:	1919 mi[1] (**unknown operator: u'strong'** km)
Existed:	1926[1] – present
Major junctions	
North end:	160 US 160 at Pagosa Springs, CO
	25 85 285 I-25 / US 85 / US 285 at Santa Fe, NM 40 54 I-40 / US 54 at Santa Rosa, NM 27 87 I-27 / US 87 at Lubbock, TX 20 I-20 at Roscoe, TX 20 I-20 at Abilene, TX 35 77 I-35 / US 77 at Waco, TX 45 I-45 at Fairfield, TX 49 I-49 near Mansfield, LA 55 I-55 near Brookhaven, MS 65 I-65 at Evergreen, AL 75 I-75 at Valdosta, GA
East end:	95 38 405 I-95 / SR 38 / SR 405 near Midway, GA
East end:	I
East end:	I
East end:	I3
East end:	I
Highway system	
United States Numbered Highways	

U.S. Route 84 is an east–west United States highway. It started as a short Georgia-Alabama route in the original 1926 scheme, but now extends all the way to Colorado. The highway's eastern terminus is a short distance east of Midway, Georgia, at an intersection with I-95. The road continues toward the nearby Atlantic Ocean as a county road. Its western terminus is in Pagosa Springs, Colorado, at an intersection with U.S. Route 160.[2]

The section from Brunswick, Georgia to Roscoe, Texas has been designated by five state legislatures as part of the El Camino East/West Corridor. The designation was in recognition of its history as a migration route from the Atlantic coast to the present U.S.-Mexico border, one of the routes that Spanish settlers called *El Camino Real*. (In Louisiana, the route was called the Harrisonburg Road.) The designation is intended to promote the route for both tourism and NAFTA-facilitated trade with Mexico.[3] [4] States are asking for Federal funds to four-lane the U.S. 84 El Camino East/West Corridor.

Route description

Colorado

The western endpoint of U.S. 84, Pagosa Springs, Colorado, was made famous by C.W. McCall in the 1975 song and album *Wolf Creek Pass*. U.S. 84 actually ends approximately 1 mile (**unknown operator: u'strong'** km) east of downtown Pagosa Springs at a T-intersection with U.S. 160 (Main Street).

South of Pagosa Springs, the 28 miles (**unknown operator: u'strong'** km) Colorado miles of U.S. 84 pass through a portion of San Juan National Forest. The highway climbs over a drainage divide between the Rio Blanco and Navajo River to the village of Chromo before passing into New Mexico.

New Mexico

U.S. 84 enters Rio Arriba County, New Mexico 28 miles (**unknown operator: u'strong'** km) south of its terminus at U.S. Route 160. About 6 miles (**unknown operator: u'strong'** km) south of the Colorado-New Mexico state line, U.S. Route 64 comes from the west and cosigns with U.S. 84 for the next 28 miles (**unknown operator: u'strong'** km). Only 3 miles (**unknown operator: u'strong'** km) east of this intersection, the co-signed route crosses the Continental Divide at Sargent Pass, elevation 7718 feet (**unknown operator: u'strong'** m) above sea level or more than 3100 feet (**unknown operator: u'strong'** m) lower than Wolf Creek Pass, the next Continental Divide highway pass to the north. Therefore, only 37 miles (**unknown operator: u'strong'** km) of U.S. 84 is located west of the Continental Divide. About 12 miles (**unknown operator: u'strong'** km) east of the intersection, U.S. 64/84 enter the town of Chama, NM. At a T-intersection, New Mexico State Highway 17 enters from the north and terminates at said intersection, while U.S. 64/84 enters from the south and west.

After heading south from Chama, U.S. 64/84 combines for about 14 miles (**unknown operator: u'strong'** km) until U.S. 64 exits U.S. 84 and heads southeast, with U.S. 84 continuing south. About 57 miles (**unknown operator: u'strong'** km) down the road, U.S. 84 is joined by U.S. Route 285 south of the small community of Chili, New Mexico. About 5 miles (**unknown operator: u'strong'** km) further, U.S. 84/285 enter the town of Espanola, NM from the north as N. Paseo de Onate St. At the south end of the town, U.S. 84/285 becomes the Santa Fe Hwy and an expressway, and about 9 miles (**unknown operator: u'strong'** km) further, U.S. 84/285 becomes a limited access freeway. 15 miles (**unknown operator: u'strong'** km) further south, the two return to surface street status and then travel through downtown Santa Fe, NM. On the south side of Santa Fe at Interstate 25's exit 282A, U.S. 84/285 enter eastbound Interstate 25/U.S. Route 85. All four roads then head due south to avoid the Sangre De Cristo Mountains. Just before turning north, U.S. 285 exits the freeway at exit 290 and heads due south. After winding north and south, the freeway finally begins heading solely north, and U.S. 84 exits about 55 miles (**unknown operator: u'strong'** km) later at exit 339 near Romeroville, NM and heads in an east/southeastern direction, while Interstate 25 and U.S. 85 continue north to Colorado. About 50 miles (**unknown operator: u'strong'** km) down its solitary stretch of road, U.S. 84 enters Interstate 40 (and Historic U.S. Route 66) at Interstate 40's exit 256. Interstate 40/U.S. 84 travel through the town of Santa Rosa, NM about 23 miles (**unknown operator: u'strong'** km) later. About 25 miles (**unknown operator: u'strong'** km) from its entrance into cosigning with Interstate 40, U.S. 84 exits at exit 227.

U.S. 84 then travels south/southeast for about 53 miles (**unknown operator: u'strong'** km) until picking up U.S. Route 60 in downtown Fort Sumner, New Mexico. U.S. 84 then heads in an almost due easterly direction, and about 65 miles (**unknown operator: u'strong'** km) due east finds itself in downtown Clovis, New Mexico, having passed through the towns of Taiban, New Mexico and Melrose, New Mexico. In downtown Clovis, U.S. 60/84 is joined by U.S. Route 70.About 10 miles (**unknown operator: u'strong'** km) east, U.S. 60/70/84 pass through the center of the towns of Texico, New Mexico and Farwell, Texas, as well as over the Texas-New Mexico state line.[5]

Texas

US 84/60/70 crosses into Texas at Farwell. After passing through Farwell, US 60 makes a turn to the northeast, while US 84/70 veers to the southeast, continuing as a co-signed route until Muleshoe.

From Muleshoe, US 70 leaves the route, while US 84 continues on a southeasterly bearing across the level plains of the Llano Estacado. Along this stretch, US 84 runs parallel to the BNSF Railway, crosses a sandy section called the Muleshoe Dunes, and then passes Littlefield, Texas, the birthplace of country singer Waylon Jennings. US 84 continues in a southeasterly direction through cotton fields and small towns such as Anton and Shallowater, eventually entering Lubbock, the largest city in the South Plains and the birthplace of Buddy Holly. Signed as Avenue Q, US 84 passes through the heart of downtown Lubbock before making a sharp easterly turn on the southeast side of the city, where it is known as the Slaton Highway. After bypassing the town of Slaton, US 84 makes another gentle turn to the east, following a generally southeasterly heading through Post, Snyder, and Roscoe, where it merges with I-20.

BNSF freight train running parallel to U.S. Route 84 while crossing the Llano Estacado.

U.S. Route 84 is known in Lubbock, Texas, as Avenue Q, a major downtown thoroughfare.

From this point, US 84 follows I-20, unsigned, until Abilene where it leaves the interstate, making a hard southerly turn and forming the western side of a 3/4 loop around the city (along with US Highways 83 and 277). From the south side of Abilene, US 84 continues as a co-signed route with US 83 (signed as US 84 West/East and US 83 South/North) until the two highways split approximately 2 miles (**unknown operator: u'strong'** km) northeast of Tuscola, and, though still signed as east/west, maintains a due south/north heading. US 84 makes a gentle turn back to the southeast at Lawn, following this bearing until Santa Anna, where it merges with US 67 and takes a more due easterly turn.

US 84 sign north of Snyder, Texas

US 84 merges with US 183 at Brownwood, and once again turns to the southeast, continuing as a co-signed route until Goldthwaite, where it leaves US 283 and yet again makes a sharp turn to the east. It follows this heading all the way to McGregor. From McGregor, the highway makes a turn to the northeast to Woodway; this stretch of US 84 is also signed as the **George W. Bush Parkway**. US 84 then crosses into Waco, passing through the heart of downtown as Waco Drive, and then northeast into the suburb of Bellmead. After briefly being co-signed with State Highway 31 through Bellmead, US 84 continues more or less due east until Teague, where it takes yet another turn to the north before turning back to the east at an intersection with I-45 in Fairfield.

US 84 merges with US 79 and makes another northerly turn southwest of Palestine, and then splits from Highway 79 just southwest of downtown before making another turn eastward and passing through town. The highway follows a gentle northeasterly path all the way to Timpson, passing through the towns of Maydelle, Rusk where it intersects with US 69, Reklaw, Mount Enterprise where it intersects US 259 and Timpson where it merges with US 59, and serves as the northern terminus of SH 87 . From Timpson to Tenaha it is briefly co-signed with US 59 to its intersection with US 96. From this point, US 84 continues its easterly path through to rest of eastern Texas, passing through Joaquin before crossing into Louisiana across the Sabine River into the town of Logansport.

Louisiana

Once the highway leaves Logansport, it travels through Stanley and then northeast into Mansfield, where it merges for a brief stretch with US-171. It continues east, crossing under I-49, until it reaches Grand Bayou, where it turns to the southeast, merging with LA-1. After two miles, it turns back again to the east, where it merges with US-71 in Coushatta and stays with that highway until Clarence. It then heads northeast towards Winnfield, where it merges for a short time with US-167. It turns to the northeast towards Joyce, then begins a long stretch to the southeast, passing through Tullos, where it intersects US-165, through the parish seat of Jena, and continues in that direction until it crosses into Catahoula Parish. It bears east through Jonesville until Ferriday,

Abandoned section of U.S. 84 east of Jena, Louisiana

where it merges with both US-65 and US-425. It then heads southeast through Vidalia where it crosses the Mississippi River into Natchez.

Mississippi

The four-laned Natchez-Vidalia Bridge, crossing the Mississippi River, carries Highway 84 into Natchez here it merges with U.S. 61. It then travels 4 miles to the northeast where it reaches the western terminus of US-98 at Washington, where it is paired with US 98 until Bude/Meadville. The road continues east, crossing under I-55 and heads east towards Collins.

A new bypass of Collins opened in March 2009, relieving the heavily-congested original route through downtown. The new bypass provides improved access to U.S. Route 49, another of Mississippi's major four-lane highways connecting Jackson and points north through the Delta with the Mississippi Gulf Coast.[6] Highway 84 runs concurrent with Interstate 59 for a short distance through Laurel, including a segment currently under reconstruction to remove a notorious 'S' curve. It then heads east to Waynesboro and continues to the Alabama state line.

Alabama

In Alabama, U.S. 84 is paired with unsigned State Route 12. Parts of the route have been widened in recent years to four lane status. The most significant exist from River Falls eastward to Andalusia, near Opp where a recent bypass of the downtown area now carries US-84 in a southern arc around the town, and from Enterprise eastward thru the Dothan area and onto the Georgia state line at the Chattahoochee River.

Plans exist of widening the US-84 corridor to four lanes elsewhere in Alabama but due to funding issues the projects are in various stages of development. One such stretch is just east of Opp where construction began on a four lane eastward towards Elba but has stalled due to funding.

US-84 has brief stretches of being co-signed with other US Highways in Alabama. US 84 shares a routing with US-31 from just southwest of Evergreen to a few miles east of the town. It briefly shares a routing with US-29 in the city of Andalusia. It also shares a routing with US-331 near the city of Opp. At Dothan US-84 shares a concurrency with US-231 and US-431 on the Ross Clark Circle which is the circular bypass of downtown Dothan.

US-84 crosses three major navigable waterways in Alabama. They are the Tombigbee River at Coffeeville, the Alabama River at Claiborne, and the Chattahoochee River at the Alabama-Georgia state line. US-84 passes very near the large military facility of Fort Rucker which is the US Army's helicopter training school. Most of US-84 in Alabama traverses the Gulf Coastal Plain which is relatively lowlying land with some sand content. The area is heavily agricultural with little heavy industry. Dothan is the largest city in Alabama traversed by US-84 and it is the business and agricultural center of Southeast Alabama. The area is commonly referred to as the Wiregrass Region.

History

The original 1926 route of U.S. 84 skirted the southern border of Georgia, from Brunswick to the north edge of the Okefenokee Swamp, then west to Dothan, Alabama, just across the Alabama line.

In 1934, U.S. 84 was extended to Grove Hill, Alabama, then south on US 43 to Wagarville, Alabama, west to State Line, Mississippi, north on US 45 to Waynesboro, Mississippi, and then across Mississippi and Louisiana to Farwell, Texas. State Line was bypassed in the 1960s by a direct connection between Grove Hill and Waynesboro. A few sources report that the part between Natchez, Mississippi and Wagarville was planned as **US 86** a year before. The Alabama Department of Transportation library in Montgomery, Alabama, holds state-issued maps and documents from that era with the stretch from US 43 to Mississippi labeled that way. At one point, funding was not secure for building a bridge over the Alabama River, and a US 86 designation would have made the absence of a bridge less obvious.[7]

The east ends of U.S. 84 and U.S. Route 82 were swapped in 1989 after the roads around Waycross, Georgia, were reconfigured.

References

[1] US Highways from US 1 to US 830 (http://www.us-highways.com/us1830.htm) Robert V. Droz

[2] Endpoints of U.S. 84 (http://usends.com/80-89/084/084.html)

[3] El Camino East/West Commission website (http://elcaminocorridor.org/)

[4] Alabama's Joint Resolution about the El Camino East/West Corridor (http://www.legislature.state.al.us/SearchableInstruments/2004RS/Resolutions/HJR146.htm)

[5] Google Maps ©2009 , Tele Atlas

[6] http://www.gomdot.com/Home/MediaRoom/newsreleases/PressReleaseDetail.aspx?ID=32200995618

[7] Endpoints of U.S. 86 (http://www.webcitation.org/query?url=http://www.geocities.com/mapguy_annex/HwyEnds/End086/end086.htm&date=2009-10-26+02:11:40)

Browse numbered routes

Alabama State Route 27

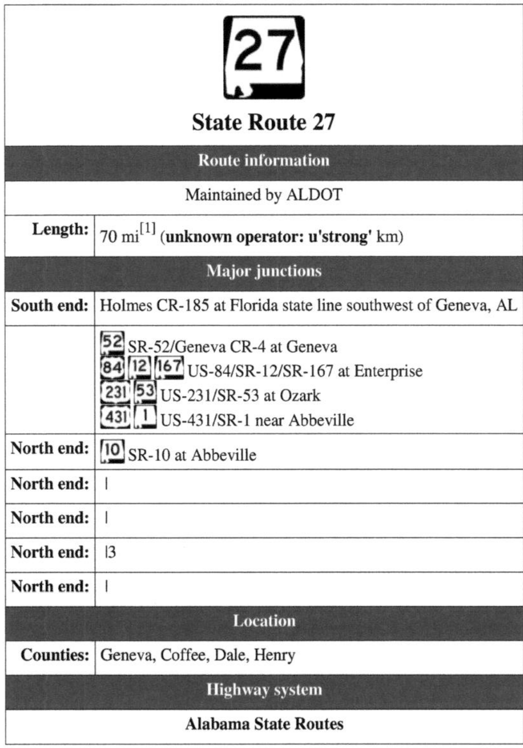

State Route 27

Route information	
Maintained by ALDOT	
Length:	70 mi[1] (**unknown operator: u'strong'** km)
Major junctions	
South end:	Holmes CR-185 at Florida state line southwest of Geneva, AL
	SR-52/Geneva CR-4 at Geneva US-84/SR-12/SR-167 at Enterprise US-231/SR-53 at Ozark US-431/SR-1 near Abbeville
North end:	SR-10 at Abbeville
North end:	I
North end:	I
North end:	I3
North end:	I
Location	
Counties:	Geneva, Coffee, Dale, Henry
Highway system	
Alabama State Routes	

State Route 27 is a 70-mile (**unknown operator: u'strong'** km) long route in the southeastern part of the state. The southern terminus of the route is at the Florida-Alabama state line, where the route serves as a continuation of Holmes County Road 185. The northern terminus of the route is at the junction with State Route 10 at Abbeville.

Route description

Entering Alabama in eastern Geneva County, State Route 27 is primarily a two-lane route. It serves as a connector from the Gulf coast beaches along the Florida Panhandle to major routes such as U.S. Highway 84 and U.S. Highway 231. Between the cities of Geneva and Enterprise, the route's orientation is generally north–south. Upon leaving Enterprise, the route assumes a northeastward trajectory as it leads into Dale County. East of Ozark, the route, while still signed as "north" and "south", assumes an east–west orientation until its terminus at Abbeville

Cities along the route

- Geneva
- Enterprise
- Ozark
- Abbeville

Major intersections

- State Route 196 at Geneva
- State Route 52 at Geneva
- State Route 85 north of Geneva
- State Route 192 at Central City
- State Route 134 at Enterprise
- State Route 88 at Enterprise
- U.S. Highway 84 at Enterprise
- State Route 248 at Enterprise
- U.S. Highway 231 at Ozark
- State Route 249 at Ozark
- State Route 105 at Ozark
- State Route 173 southeast of Abbeville
- U.S. Highway 431 south of Abbeville
- State Route 95 at Abbeville

References

[1] Highway Route Info Routes 26 - 50 (http://colepages.com/routes2650.html)

Newton, Alabama

<table>
<tr><td colspan="2" align="center">Newton, Alabama</td></tr>
<tr><td colspan="2" align="center">— Town —</td></tr>
<tr><td colspan="2" align="center">
Downtown Newton, Alabama</td></tr>
<tr><td colspan="2" align="center">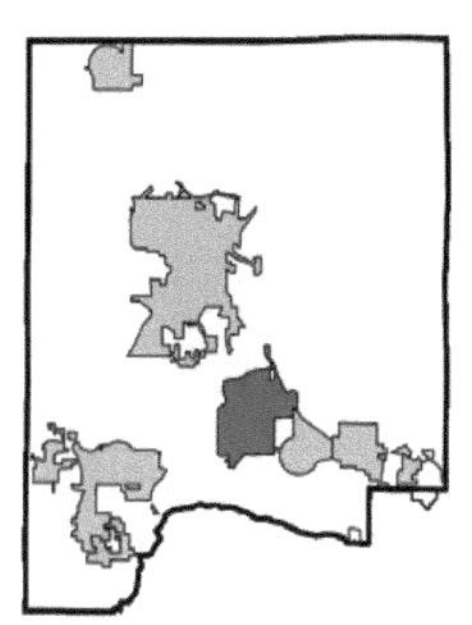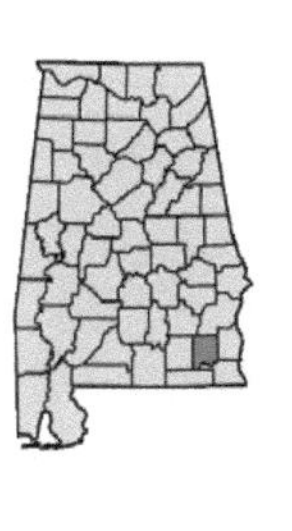
Location in Dale County and the state of Alabama</td></tr>
<tr><td colspan="2" align="center">Coordinates: 31°20′40″N 85°35′34″W</td></tr>
<tr><td>Country</td><td>United States</td></tr>
<tr><td>State</td><td>Alabama</td></tr>
<tr><td>County</td><td>Dale</td></tr>
<tr><td>Area</td><td></td></tr>
<tr><td>• Total</td><td>14.3 sq mi (37.1 km^2)</td></tr>
<tr><td>• Land</td><td>14.3 sq mi (37 km^2)</td></tr>
<tr><td>• Water</td><td>0 sq mi (0.1 km^2)</td></tr>
<tr><td>Elevation</td><td>318 ft (97 m)</td></tr>
</table>

Population (2000)	
• Total	1708
• Density	119.4/sq mi (46/km^2)
Time zone	Central (CST) (UTC-6)
• Summer (DST)	CDT (UTC-5)
ZIP code	36352
Area code(s)	334
FIPS code	01-54480
GNIS feature ID	0123931

Newton is a town in Dale County, Alabama, United States. At the 2000 census its population was 1,708. Once the county seat of Dale County, Newton lost this distinction to nearby Ozark in 1870, and is now a small farming community. The city currently forms a part of the Enterprise-Ozark micropolitan statistical area.

Geography

Newton is located at 31°20′40″N 85°35′34″W.[1]

According to the U.S. Census Bureau, the town has a total area of 14.3 square miles (**unknown operator: u'strong'** km^2), of which, 14.3 square miles (**unknown operator: u'strong'** km^2) of it is land and 0.04 square miles (**unknown operator: u'strong'** km^2) of it (0.21%) is water.

The Choctawhatchee River flows just to the north and west of Newton. At one time this waterway was navigable by steamboat from its mouth in Choctawhatchee Bay, Florida all the way to Newton.[2]

Alabama State Route 123 and Alabama State Route 134 both pass through Newton.

History

Civil War Monument in Newton, Alabama

Newton was founded in 1843 after the formation of Coffee County from Dale County's western half, which rendered the original county seat of Daleville off-center. The town became a scene for local Confederate recruiting during the Civil War, and was the site of a battle in March 1865 between local Home Guard units and pro-Union irregulars operating out of Florida. The guerrillas, who sought to burn the county courthouse, were repulsed. This event is commemorated by a monument located in downtown Newton, and by periodic re-enactments.[3]

On December 3, 1864, a local Methodist minister named Bill Sketoe was lynched just north of Newton by local Home Guard elements led by Captain Joseph Brear. Since Sketoe was tall, a hole had to be dug beneath his feet to accommodate his large frame. Local legend insists that "the hole that won't stay filled" never vanished—even after being filled in numerous times during the decades that followed. Though covered in 1979 by a new bridge and tons of rip-rap, "Sketoe's hole" remains a local attraction,[4] and was immortalized by Alabama writer Kathryn Tucker Windham in her 13 Alabama Ghosts and Jeffrey.[5] A monument to Sketoe was dedicated near the hanging site in 2006,[6] and the local museum displays items of Sketoe memoribilia.[7]

Following a fire which destroyed the courthouse in March 1869, and the formation of Geneva County in 1870 from the southern third of Dale and Coffee Counties, voters relocated the county seat to Ozark, which was more centralized.

The Southern Star, one of the oldest newspapers in the Wiregrass area, was first published in Newton in 1867. It later relocated to Ozark, where it continues to be published today. Newton remained a port for river boats on the nearby Choctawhatchee, until the railroad arrived in 1890. The Baptist Collegiate Institute operated in the city from 1898 to 1929; its main building now houses the city's public library.

Demographics

As of the census[8] of 2000, there were 1,708 people, 693 households, and 510 families residing in the town. The population density was 119.5 people per square mile (46.1/km²). There were 790 housing units at an average density of 55.3 per square mile (21.3/km²). The racial makeup of the town was 82.20% White, 15.52% Black or African American, 0.18% Native American, 0.06% Asian, 0.06% Pacific Islander, 0.06% from other races, and 1.93% from two or more races. 1.23% of the population were Hispanic or Latino of any race.

There were 693 households out of which 31.7% had children under the age of 18 living with them, 56.6% were married couples living together, 13.3% had a female householder with no husband present, and 26.4% were non-families. 24.1% of all households were made up of individuals and 9.4% had someone living alone who was 65 years of age or older. The average household size was 2.46 and the average family size was 2.93.

In the town the population was spread out with 24.0% under the age of 18, 8.9% from 18 to 24, 27.2% from 25 to 44, 25.8% from 45 to 64, and 14.1% who were 65 years of age or older. The median age was 39 years. For every 100 females there were 93.9 males. For every 100 females age 18 and over, there were 89.8 males.

The median income for a household in the town was $33,021, and the median income for a family was $35,795. Males had a median income of $28,924 versus $19,559 for females. The per capita income for the town was $15,263. About 13.3% of families and 13.1% of the population were below the poverty line, including 17.8% of those under age 18 and 8.5% of those age 65 or over.

Schools

Newton has its own Elementary School [9].

References

[1] "US Gazetteer files: 2010, 2000, and 1990" (http://www.census.gov/geo/www/gazetteer/gazette.html). United States Census Bureau. 2011-02-12. . Retrieved 2011-04-23.

[2] Wiregrass Saga (http://www.southern-style.com/Southeast Alabama Heritage Association/Oscar Tompkins Wire Grass Sagas.htm), by Oscar L. Thompson. Retrieved on 2009-04-30.

[3] Civil War Groups to Bring History to Life in Newton (http://www.zwire.com/news/newsstory.cfm?newsid=20167799&title=Civil+War+groups+to+bring+history+to+life+in+Newton&BRD=1145&PAG=461&CATNAME=Daleville/Dale+News&CATEGORYID=418). Retrieved on 2009-04-30.

[4] The Hanging of Bill Sketoe (http://homepage.mac.com/katwhiskers/iblog/B1938900754/C871611634/E1619713658/index.html). Retrieved on 2009-04-30.

[5] Windham, Katheryn Tucker: *13 Alabama Ghosts and Jeffrey*. Strode Publishers, 1969. ISBN 978-0-8173-0376-1.

[6] Sketoe Tale Memorialized (http://www.zwire.com/news/newsstory.cfm?newsid=17383910&title=Sketoe+tale+memorialized&BRD=1145&PAG=461&CATNAME=Top+Stories&CATEGORYID=410). Retrieved on 2009-04-30.

[7] Historic Building Serves New Purpose as Town Hall (http://www.zwire.com/news/newsstory.cfm?newsid=20127513&title=Historic+building+serves+new+purpose+as+town+hall&BRD=1145&PAG=461&CATNAME=Daleville/Dale+News&CATEGORYID=418). Retrieved on 2009-05-01.

[8] "American FactFinder" (http://factfinder.census.gov). United States Census Bureau. . Retrieved 2008-01-31.

[9] http://dalecountyboe.org/nes.htm

Ozark, Alabama

<table>
<tr><td colspan="2" align="center">Ozark, Alabama
— City —</td></tr>
<tr><td colspan="2" align="center">Location in Dale County and the state of Alabama</td></tr>
<tr><td colspan="2" align="center">Coordinates: 31°26′53″N 85°38′31″W</td></tr>
<tr><td>Country</td><td>United States</td></tr>
<tr><td>State</td><td>Alabama</td></tr>
<tr><td>County</td><td>Dale</td></tr>
<tr><td>Settled</td><td></td></tr>
<tr><td>Incorporated</td><td>October 27, 1870</td></tr>
<tr><td>Government</td><td></td></tr>
<tr><td>• Type</td><td>Mayor/City Council</td></tr>
<tr><td>• Mayor</td><td>Billy Blackwell</td></tr>
<tr><td>Area</td><td></td></tr>
<tr><td>• Total</td><td>34.4 sq mi (89.3 km^2)</td></tr>
<tr><td>• Land</td><td>34.2 sq mi (88.7 km^2)</td></tr>
<tr><td>• Water</td><td>0.2 sq mi (0.6 km^2)</td></tr>
<tr><td>Elevation</td><td>417 ft (127 m)</td></tr>
<tr><td>Population (2000)</td><td></td></tr>
<tr><td>• Total</td><td>15119</td></tr>
<tr><td>• Density</td><td>439.5/sq mi (169.3/km^2)</td></tr>
<tr><td>Time zone</td><td>Central (CST) (UTC-6)</td></tr>
<tr><td>• Summer (DST)</td><td>CDT (UTC-5)</td></tr>
<tr><td>ZIP codes</td><td>36360-36361</td></tr>
<tr><td>Area code(s)</td><td>334</td></tr>
</table>

FIPS code	01-57648
GNIS feature ID	0152832
Website	http://www.ozarkalabama.us/

Ozark is a city in and the county seat of Dale County, Alabama, United States.[1] In 1990 the city population was 12,922.[2] As of the 2000 census, the population of the city was 15,119.[3]

Ozark along with the city of Enterprise, are principal cities of the Enterprise–Ozark Micropolitan Statistical Area, which includes all of Dale and Coffee counties. Fort Rucker, the primary flight training base for Army Aviation, abuts Ozark.

History

The Ozark area was originally inhabited by the Creek Indians. It is said that Ozark received its name after a traveller thought it reminded him of the Ozark Mountains in Arkansas.

The first known European settler in Ozark was John Merrick Sr., a veteran of the Revolutionary War, in 1822. In honor of him, the town was named Merricks. It was later changed to Woodshop, which was its name when the town received its post office. The first appearance of the name Ozark was in 1855, when the citizens requested a name change.[4]

The county seat was moved from Newtown to Ozark 1870.[4] [5]

Geography

Ozark is located at 31°26′53″N 85°38′31″W (31.448169, -85.642009)[6] . It is part of the Wiregrass Region.

According to the U.S. Census Bureau, the city has a total area of 34.5 square miles (**unknown operator: u'strong'** km^2) of which 34.2 square miles (**unknown operator: u'strong'** km^2) is land and 0.2 square miles (**unknown operator: u'strong'** km^2) (0.70%) is water.[7]

Demographics

As of the census[8] of 2000, there were 15,119 people, 6,126 households, and 4,233 families residing in the city.[3] The population density was 441.5 inhabitants per square mile (**unknown operator: u'strong'** /km^2). There were 6,955 housing units at an average density of 203.1 per square mile (**unknown operator: u'strong'** /km^2).[7] The racial makeup of the city was 68.28% White, 28.30% Black or African American, 0.67% Native American, 0.70% Asian, 0.05% Pacific Islander, 0.46% from other races, and 1.53% from two or more races. 2.08% of the population were Hispanic or Latino of any race.[3]

There were 6,126 households out of which 31.4% had children under the age of 18 living with them, 48.8% were married couples living together, 16.9% had a female householder with no husband present, and 30.9% were non-families. 27.7% of all households were made up of individuals and 10.6% had someone living alone who was 65 years of age or older. The average household size was 2.40 and the average family size was 2.92.[3]

In the city the population was spread out with 24.9% under the age of 18, 8.5% from 18 to 24, 26.0% from 25 to 44, 24.7% from 45 to 64, and 15.9% who were 65 years of age or older. The median age was 39 years. For every 100 females there were 87.8 males. For every 100 females age 18 and over, there were 83.4 males.[3]

The median income for a household in the city was $29,330, and the median income for a family was $38,633. Males had a median income of $30,236 versus $19,564 for females. The per capita income for the city was $15,984. About 14.8% of families and 19.0% of the population were below the poverty line, including 27.0% of those under age 18 and 18.0% of those age 65 or over.[3]

Education

Ozark is served by the Ozark City Schools. Schools located in the city are Carroll High School (grades 9 through 12), Carroll Career Center (grades 9 through 12), D.A. Smith Middle School (grades 6 through 8), Harry M. Mixon Elementary School (grades 1 through 5), Joseph W. Lisenby Elementary School (grades 1 through 5) and Mamie Thompkins School (Kindergarten).

There are three private schools in Ozark - Harvest Christian School, Dale County Christian School, and Eastgate Christian Academy (all grades K through 12).

Post-secondary education is available at Enterprise State Community College's Alabama Aviation Center at Ozark. Programs are offered in in Aviation maintenance technology.

Carroll High School Band 2009

Carroll High School Marching Band "Pride of the Wiregrass" 2007

Media

Radio stations

- WAQG 91.7 FM (Contemporary Christian)
- WJRL 103.9 FM (Sports talk)
- WOAB 104.9 FM (Adult standards
- WOZK 900 AM (MOR)
- WQLS 1210 AM (Gospel)

Newspaper

- *The Southern Star*- weekly

Notable residents

- Wilbur Jackson – National Football League
- Byron Mitchell - former super middleweight boxing champion.
- Steve McLendon - Nose Tackle/Defensive End, Pittsburgh Steelers

See also

- Ozark Civic Center
- Ozark City Schools

References

[1] "Find a County" (http://www.naco.org/Counties/Pages/FindACounty.aspx). National Association of Counties. . Retrieved 2011-06-07.

[2] "Fast Facts- 1990 Data Set Ozark city, Alabama" (http://factfinder.census.gov/servlet/QTTable?_bm=y&-context=qt&
-qr_name=DEC_1990_STF1_DP1&-ds_name=DEC_1990_STF1_&-CONTEXT=qt&-tree_id=100&-redoLog=true&-all_geo_types=N&
-geo_id=label&-geo_id=16000US011330&-search_results=16000US011330&-format=&-_lang=en). *DP-1. General Population and
Housing Characteristics:*. United States Census Bureau. . Retrieved 29 January 2010.

[3] "Fact Sheet- Ozark city, Alabama" (http://factfinder.census.gov/servlet/SAFFFacts?_event=Search&geo_id=&_geoContext=&
_street=&_county=ozark&_cityTown=ozark&_state=04000US01&_zip=&_lang=en&_sse=on&pctxt=fph&pgsl=010&
show_2003_tab=&redirect=Y). *American Fast Facts*. United States Census Bureau. . Retrieved 29 January 2010.

[4] Watson, Fred Shelton (1968). *Forgotten trails: a history of Dale County, Alabama, 1824-1966*. Birmingham, Alabama: Banner Press.
pp. 288.

[5] Hoskins Morton, Patricia (10 December 2010). "Dale County" (http://www.encyclopediaofalabama.org/face/Article.jsp?id=h-1327).
Encyclopedia of Alabama. . Retrieved 1 February 2010.

[6] "US Gazetteer files: 2010, 2000, and 1990" (http://www.census.gov/geo/www/gazetteer/gazette.html). United States Census Bureau.
2011-02-12. . Retrieved 2011-04-23.

[7] "Geographic Comparison Table- Alabama" (http://factfinder.census.gov/servlet/GCTTable?_bm=y&-geo_id=04000US01&
-_box_head_nbr=GCT-PH1&-ds_name=DEC_2000_SF1_U&-format=ST-7). *American Fast Facts*. United States Census Bureau. .
Retrieved 29 January 2010.

[8] "American FactFinder" (http://factfinder.census.gov). United States Census Bureau. . Retrieved 2008-01-31.

External links

- Official city website (http://www.ozarkalabama.org)

Article Sources and Contributors

Alabama State Route 123 *Source*: http://en.wikipedia.org/w/index.php?title=Alabama_State_Route_123 *Contributors*: Arondello, Fredddie, Jllm06, Patriarca12

Dale County, Alabama *Source*: http://en.wikipedia.org/w/index.php?title=Dale_County%2C_Alabama *Contributors*: AXRL, Acntx, Altairisfar, Angr, Anivron, Busheviok, CC21560, Chphe, Danny, DothanCitizen, Dufekin, Ecjmartin, Firehawk7, Fretent, Giraffedata, GrahamHardy, HennessyC, Hyperfast, Jmlk17, Johnpacklambert, Khukri, MJCdetroit, MacRusgail, NE2, Numbo3, Omnedon, Postdlf, Ram-Man, Rich Farmbrough, Template namespace initialisation script, Theduckman1763, Tim!, Timneu22, VerruckteDan, Vladsinger, Voyager, Woohookitty, Worldenc, Wwoods, Zoe, 61 anonymous edits

Houston County, Alabama *Source*: http://en.wikipedia.org/w/index.php?title=Houston_County%2C_Alabama *Contributors*: 90, AXRL, Acntx, Altairisfar, AuburnPilot, Azumanga1, Bert The Turtle, Chphe, Christopher Hollis, Computerdad333, DanMS, Danny, DothanCitizen, Dufekin, Eric outdoors, Fairlane64, Gilliam, GrahamHardy, HennessyC, Hyperfast, Ichabod, Johnpacklambert, MJCdetroit, NE2, Omnedon, Pearle, Ram-Man, Rhatsa26X, Rich Farmbrough, Rlvaughn, Rptilby, Rsj1987, Tajas1, Template namespace initialisation script, Theduckman1763, Tide rolls, Timneu22, VerruckteDan, Vladsinger, Voyager, WillC, Worldenc, Zarcadia, Zoe, Zoicon5, 25 anonymous edits

Geneva County, Alabama *Source*: http://en.wikipedia.org/w/index.php?title=Geneva_County%2C_Alabama *Contributors*: 90, AXRL, Acntx, Altairisfar, Bluez bekztra, Danny, Dufekin, Frankie816, Fretent, GcSwRhlc, GrahamHardy, HennessyC, Hokeman, Hyperfast, Hégésippe Cormier, Johnpacklambert, LeoNomis, Lindum, MJCdetroit, Marketinsights, Mjrmtg, NE2, Nyttend, Omnedon, Pearle, Ram-Man, Rich Farmbrough, Robertjohnsonrj, Ron Ritzman, Ser Amantio di Nicolao, Sgt Pinback, Tajas1, Template namespace initialisation script, Theduckman1763, Timneu22, Tnxman307, Tolea93, VerruckteDan, Vladsinger, Voyager, WillC, Worldenc, Zoe, Zoicon5, 21 anonymous edits

Alabama State Route 52 *Source*: http://en.wikipedia.org/w/index.php?title=Alabama_State_Route_52 *Contributors*: AuburnPilot, CC21560, Doug Haynes, Fredddie, Imzadi1979, NE2, Rschen7754, TwinsMetsFan, 3 anonymous edits

Alabama State Route 103 *Source*: http://en.wikipedia.org/w/index.php?title=Alabama_State_Route_103 *Contributors*: CC21560, Fredddie, Imzadi1979, LadyofShalott, NE2, Rschen7754, TwinsMetsFan, WP Editor 2011

Little Choctawhatchee River *Source*: http://en.wikipedia.org/w/index.php?title=Little_Choctawhatchee_River *Contributors*: Dina, Dravecky, JodyB, Ken Gallager, Qxz, Sophie means wisdom, Yunipo, 1 anonymous edits

Blackwell Field *Source*: http://en.wikipedia.org/w/index.php?title=Blackwell_Field *Contributors*: AuburnPilot, D6, Jllm06, Patriarca12, Zyxw, 1 anonymous edits

Alabama State Route 51 *Source*: http://en.wikipedia.org/w/index.php?title=Alabama_State_Route_51 *Contributors*: Arondello, AuburnPilot, CC21560, DennisIsMe, Fredddie, Imzadi1979, NE2, Ohconfucius, Rschen7754, TwinsMetsFan, 6 anonymous edits

U.S. Route 84 *Source*: http://en.wikipedia.org/w/index.php?title=U.S._Route_84 *Contributors*: Arkannis, Atanamir, Azumanga1, Bdj95, Bellhalla, Berberry, Billy Hathorn, Bryan Derksen, CC21560, Cleduc, Dan ad nauseam, DanTD, DandyDan2007, Dough4872, Erasmussen, Floridasand, Fredddie, GoingBatty, Hajor, Highfly3442, HowardSF, Imzadi1979, Jacobcook82, Jasgrider, John of Reading, JohnnyAlbert10, JustAGal, Kacie Jane, Khatru2, LabradorLover456, Leaflet, Lightmouse, Lothar1976, Mcdonaat, Michael Patrick, Minesweeper, Mjrmtg, Morriswa, Msgirl01, NE2, Niobrara, O, RedWolf, Rich Farmbrough, Robertb-dc, Rschen7754, SPUI, Sattmaster, Scott5114, Shan12478, StephenHart, Stratosphere, Susvolans, TwinsMetsFan, Updatehelper, WOSlinker, WillC, Xnatedawgx, 46 anonymous edits

Alabama State Route 27 *Source*: http://en.wikipedia.org/w/index.php?title=Alabama_State_Route_27 *Contributors*: CC21560, Daniel73480, Imzadi1979, Ltljltlj, NE2, TwinsMetsFan, 2 anonymous edits

Newton, Alabama *Source*: http://en.wikipedia.org/w/index.php?title=Newton%2C_Alabama *Contributors*: Altairisfar, CapitalR, Chris the speller, Damieng, DothanCitizen, Droll, Ecjmartin, Jllm06, Johnpacklambert, Kitia, Nyttend, Ram-Man, Rettetast, Rich Farmbrough, Seth Ilys, Tresiden, 3 anonymous edits

Ozark, Alabama *Source*: http://en.wikipedia.org/w/index.php?title=Ozark%2C_Alabama *Contributors*: Abductive, Acntx, AlabamaGuy2007, Altairisfar, Arondello, Autiger, BD2412, Brain Rodeo, Can't sleep, clown will eat me, CapitalR, Cmr08, Cryptic C62, Damieng, DothanCitizen, Dr. WAPS, Ermanon, Flobster, Gaius Cornelius, Gressang, Honkywill, Jllm06, Jmlk17, John K, Johnpacklambert, Kitia, Koreyx, Mactones, Mgmizell, Mild Bill Hiccup, Motley Crue Rocks, Nyttend, Paintballerfromal, Pineboxeditor, Precious Roy, RL0919, Ram-Man, Rettetast, Rich Farmbrough, SchuminWeb, Seth Ilys, Tibetan Prayer, Trusilver, Tweety21, Wkharrisjr, 41 anonymous edits

Image Sources, Licenses and Contributors